MW00679163

Science Lab Journal

Quadrille-Ruled Pages and References

for Science Scholars

CLASSICALCONVERSATIONS.COM

Science Lab Journal: Quadrille-Ruled Pages and References for Science Scholars

Created by Courtney Sanford and Jennifer Greenholt

Published in the U.S.A. by Classical Conversations, Inc.
P.O. Box 909
West End, NC 27376

ISBN: 978-0-9904720-5-6

For ordering information, visit www.ClassicalConversationsBooks.com.
Printed in the United States of America

Science Lab Journal
Quadrille–Ruled Pages and References for Science Scholars

What do Mark Twain, Thomas Edison, General George S. Patton, Marie Curie, Thomas Jefferson, and John D. Rockefeller have in common?

Not all of these individuals were professional scientists, but all of them used the tools of scientific inquiry to pursue their passions. As they rose in their respective fields, each of these men and women kept meticulous records of their ideas and experiments, using simple notebooks to achieve great things.

- Twain sketched characters for his novels, but he also sketched inventions that he later patented.
- Edison wrestled with names for the phonograph and wrote down ideas for an electric piano and artificial silk.
- General Patton diagrammed war strategies but also outlined stained glass windows as examples of beauty.
- Madame Curie kept detailed records from her chemical experiments—her notebooks are still radioactive!
- Jefferson kept meticulous records of weather, geography, climate, and plant and animal life. It is said he carried a compass, a level, and a thermometer in his pockets everywhere he went so that he could document accurately!
- Rockefeller jotted down detailed information when he toured his oil refineries, making notes for improving his business model.

When you practice good science, taking notes and recording your findings in a lab journal, you walk in their footsteps. Follow these guidelines to make the most of your science lab journal:

- Use a well-sharpened pencil, in case you need to erase and rewrite.
- Always start by writing your name, the date, and the name of your project or experiment at the top of the page so you can find it later.
- Be neat: Illegible handwriting may be faster, but it will slow you down later when you (or others) attempt to re-read your notes.
- Take advantage of the gridlines in the notebook when you sketch, to capture proportions accurately.
- Leave room for marginal comments on one side of the page. Some scientists draw a vertical line to separate their results (large column) from their comments and tangential ideas (small column).
- A good principle is to record more than you think you will need. Later, if you type a formal lab report, you can weed out unnecessary information.

This lab journal is designed for use in science labs, but do not be afraid to apply your skills to other "field research" as well, such as visiting the Grand Canyon, testing a recipe in the kitchen, or going scuba diving. Opportunities to practice science are everywhere.

"No one should approach the temple of science with the soul of a money changer."

—Sir Thomas Browne

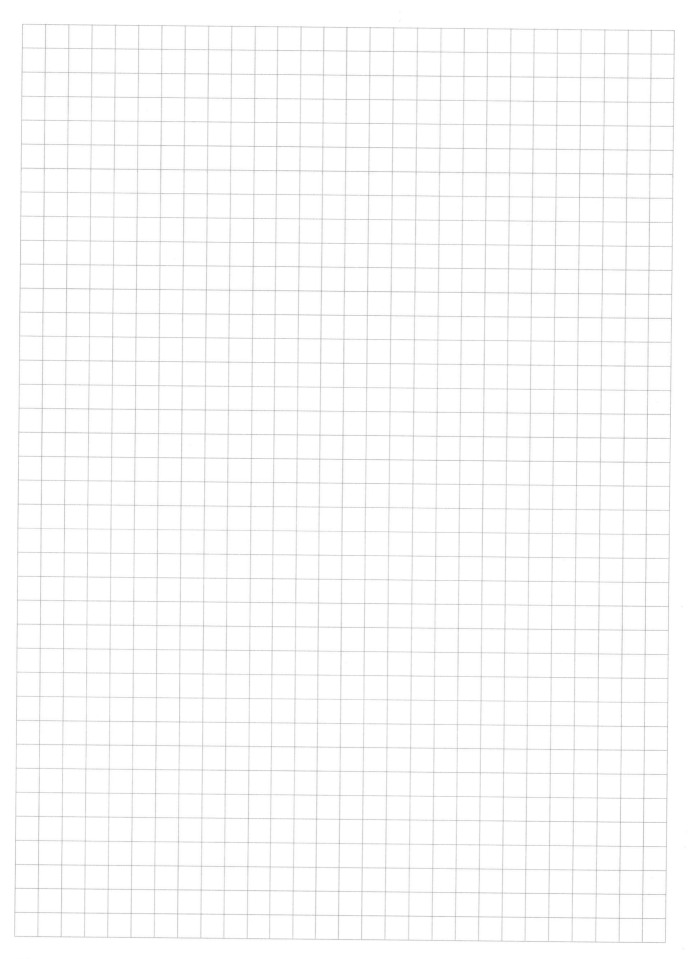

"*I* want to know how God created this world...I want to know his thoughts. The rest are details."

—Albert Einstein

24

25

26

"In all things of nature there is something of the marvelous."

—Aristotle

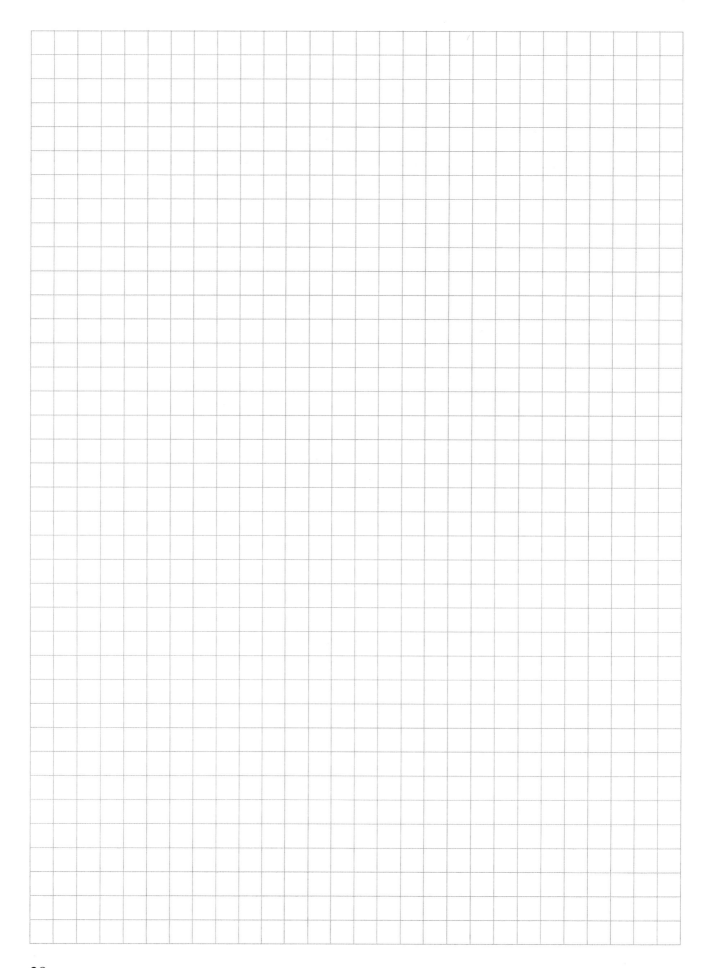

28

30

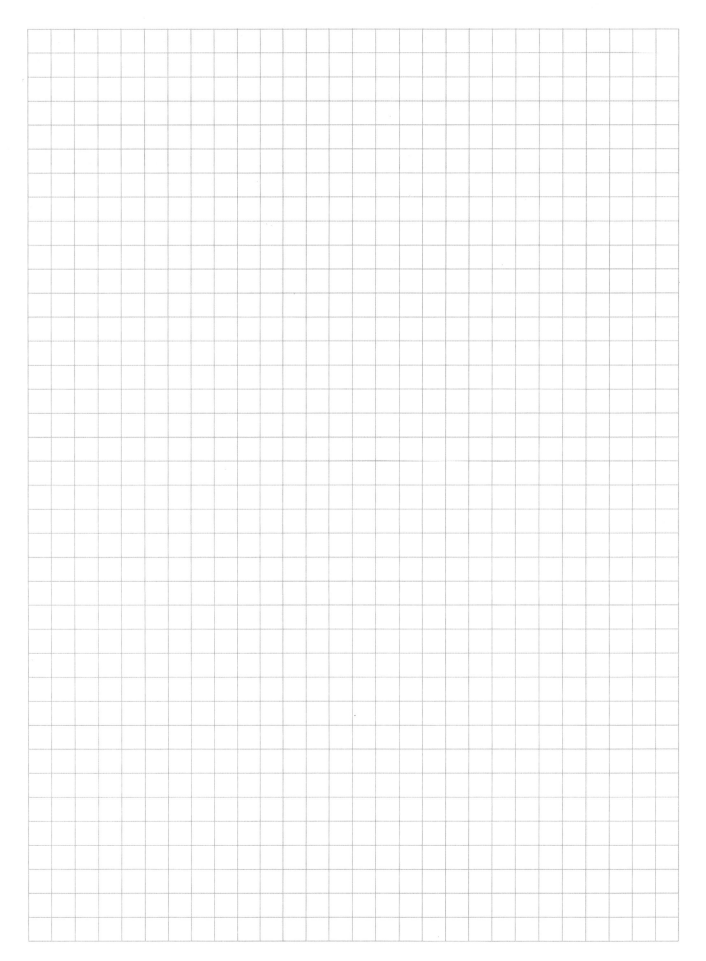

34

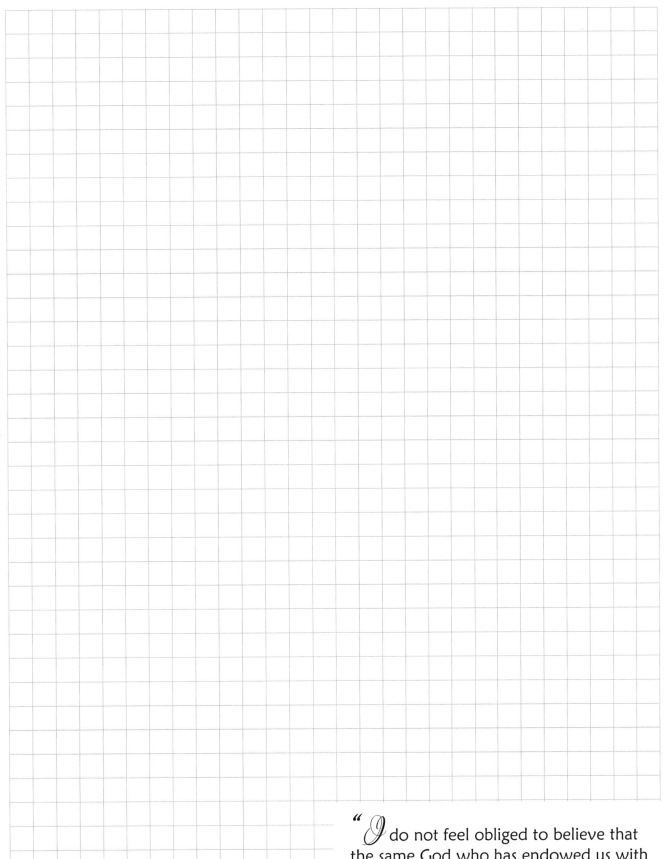

"*I* do not feel obliged to believe that the same God who has endowed us with sense, reason, and intellect has intended us to forgo their use."

—Galileo Galilei

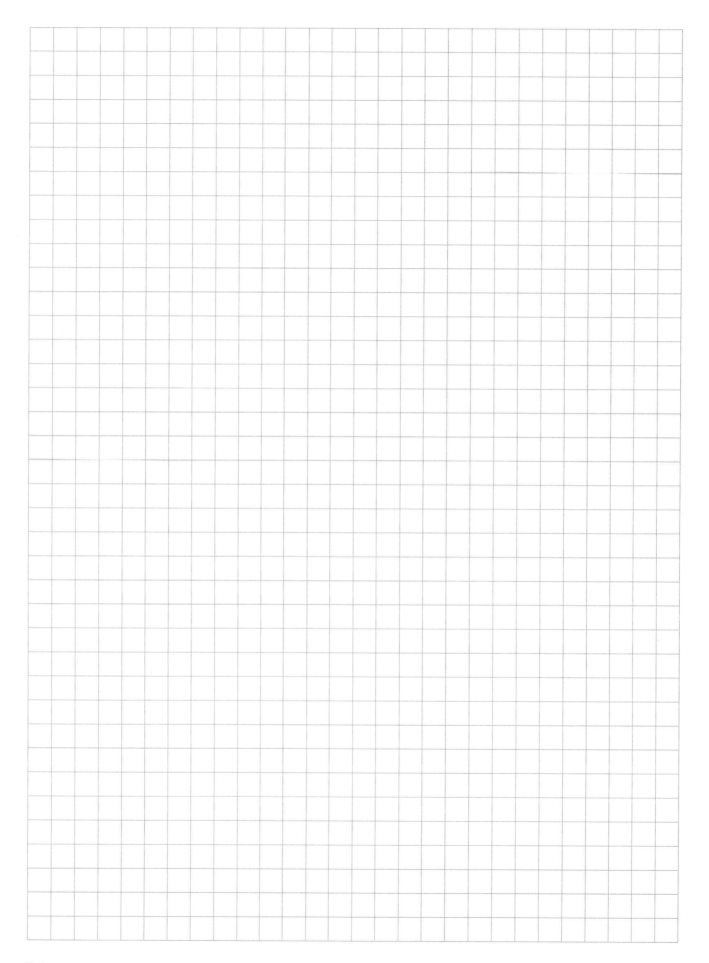

36

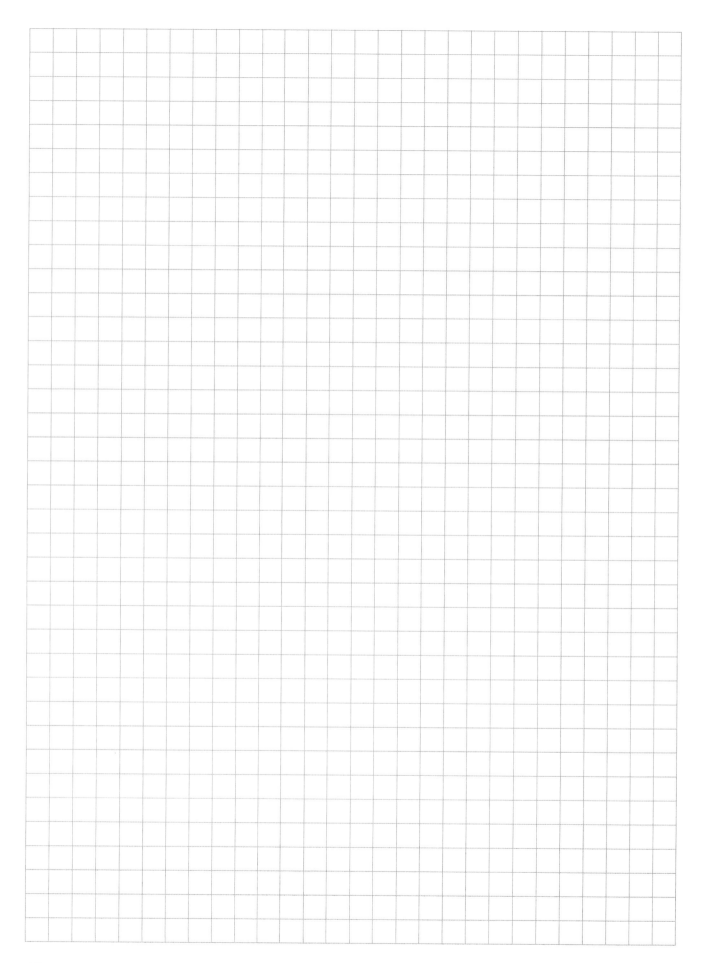

40

42

"The eternal mystery of the world is its comprehensibility.... The fact that it is comprehensible is a miracle."

—Albert Einstein

46

48

50

54

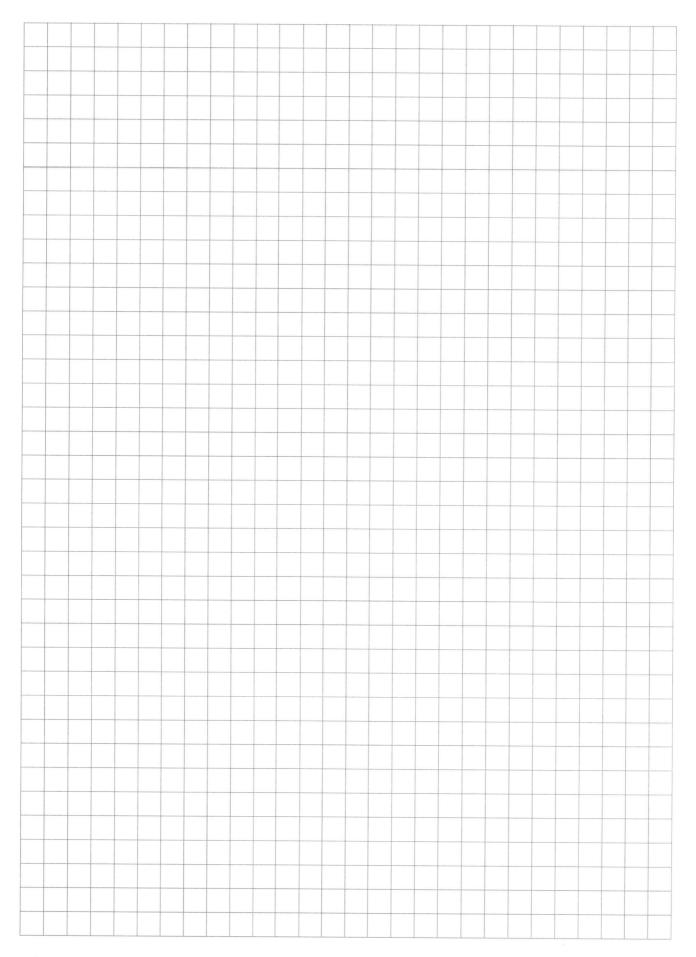

56

58

"The whole of science is nothing more than a refinement of everyday thinking."

—Albert Einstein

60

64

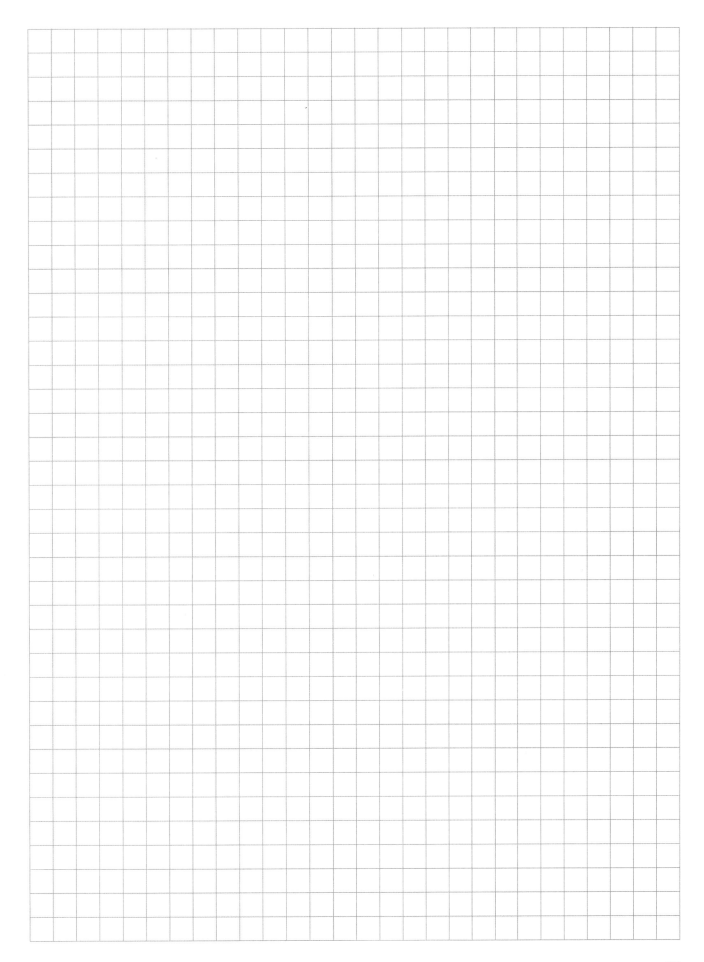

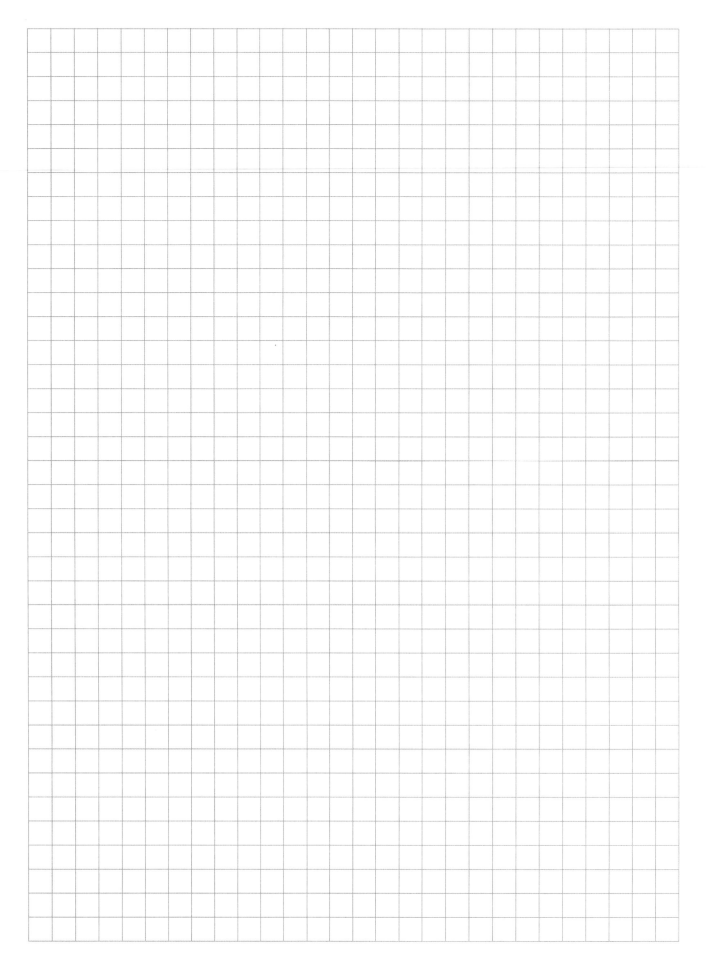

" *I*'m not smart. I try to observe. Millions saw the apple fall, but Newton was the one who asked why."

—Bernard M. Baruch

68

74

"You can observe a lot by watching."

—Yogi Berra

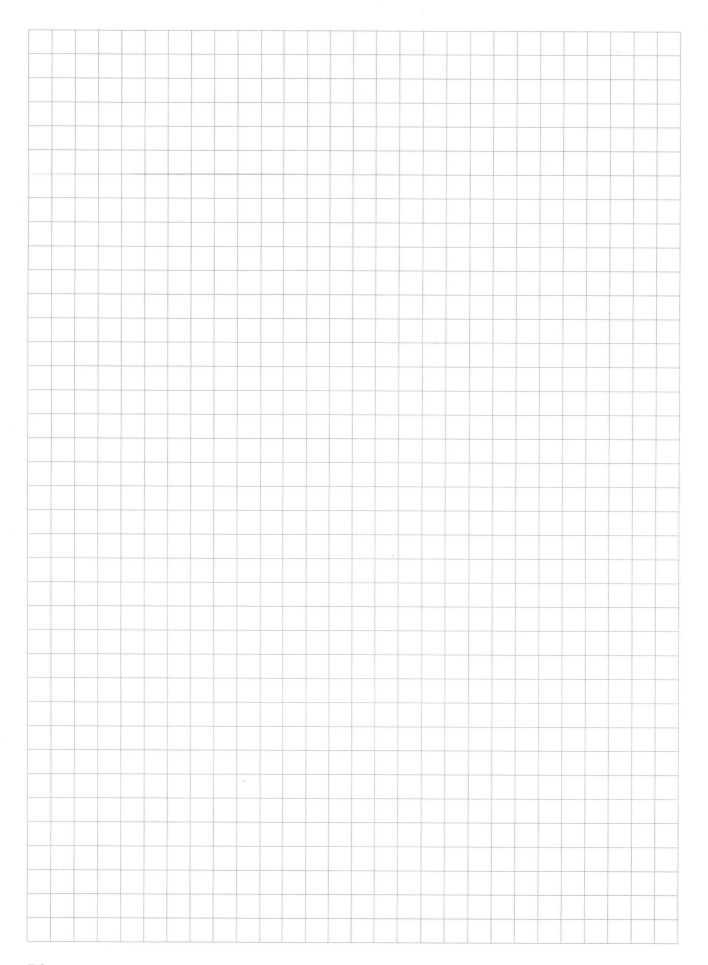

76

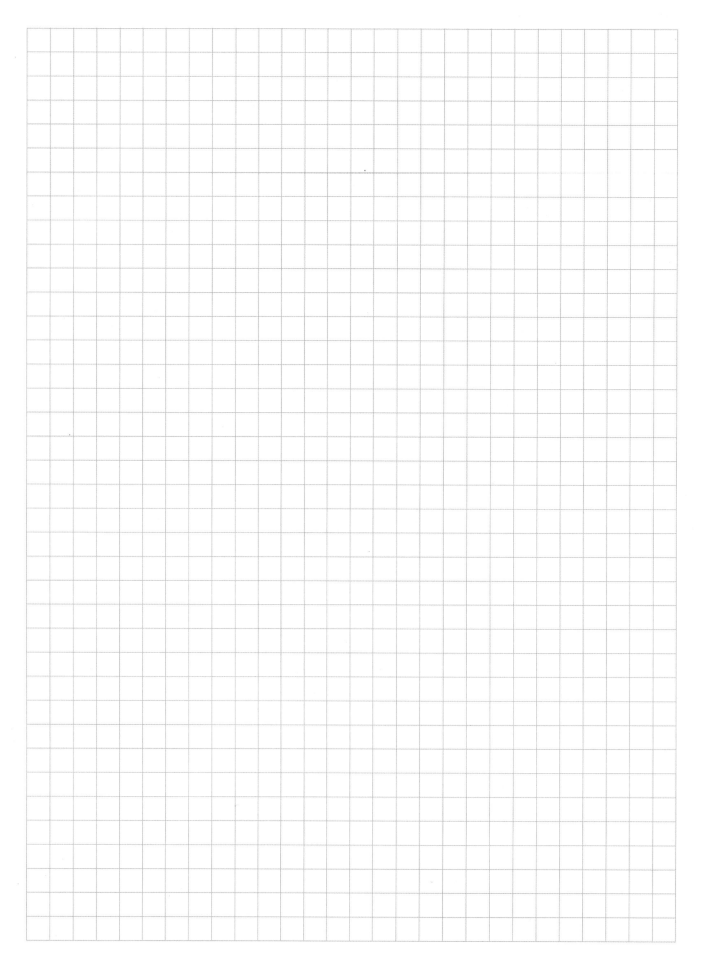

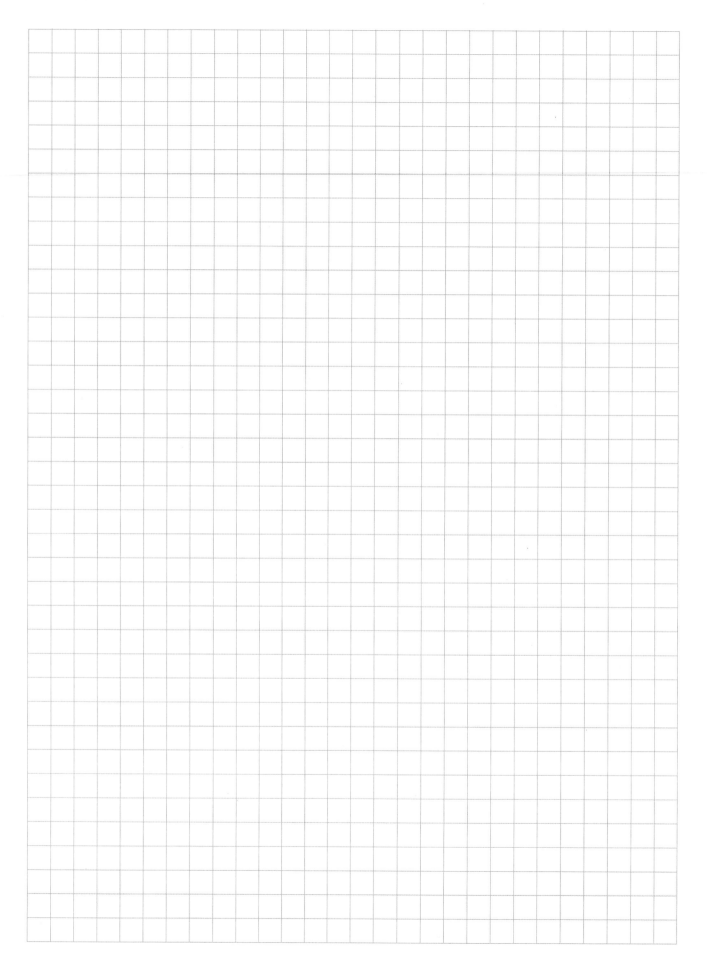

"The scientific mind does not so much provide the right answers as ask the right questions."

—Claude Lévi-Strauss

86

"The scientific mind does not so much provide the right answers as ask the right questions."

—Claude Lévi-Strauss

84

86

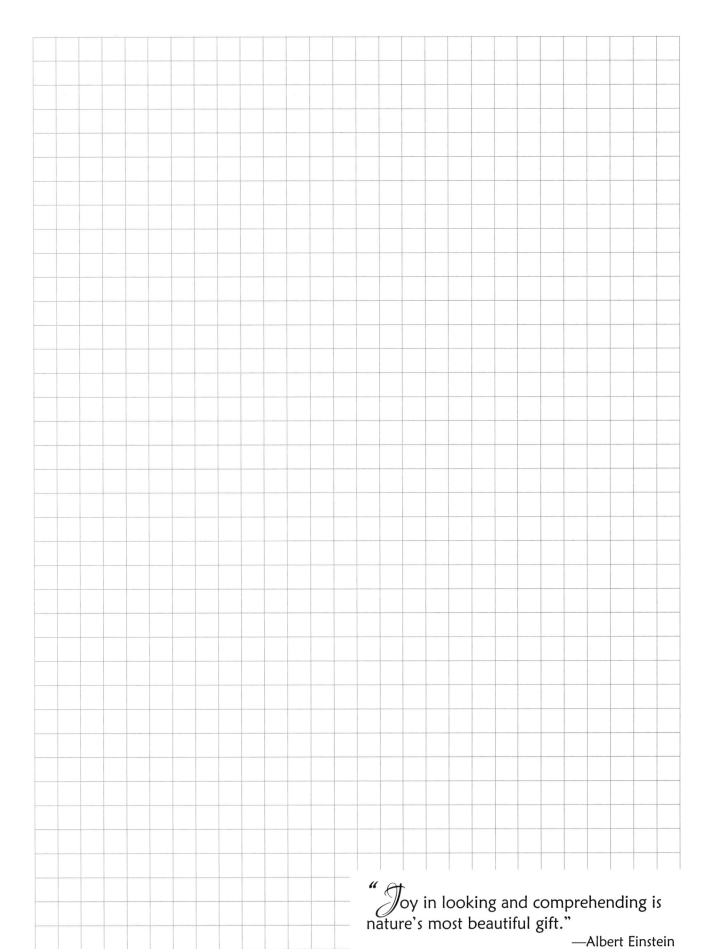

"*Joy* in looking and comprehending is nature's most beautiful gift."

—Albert Einstein

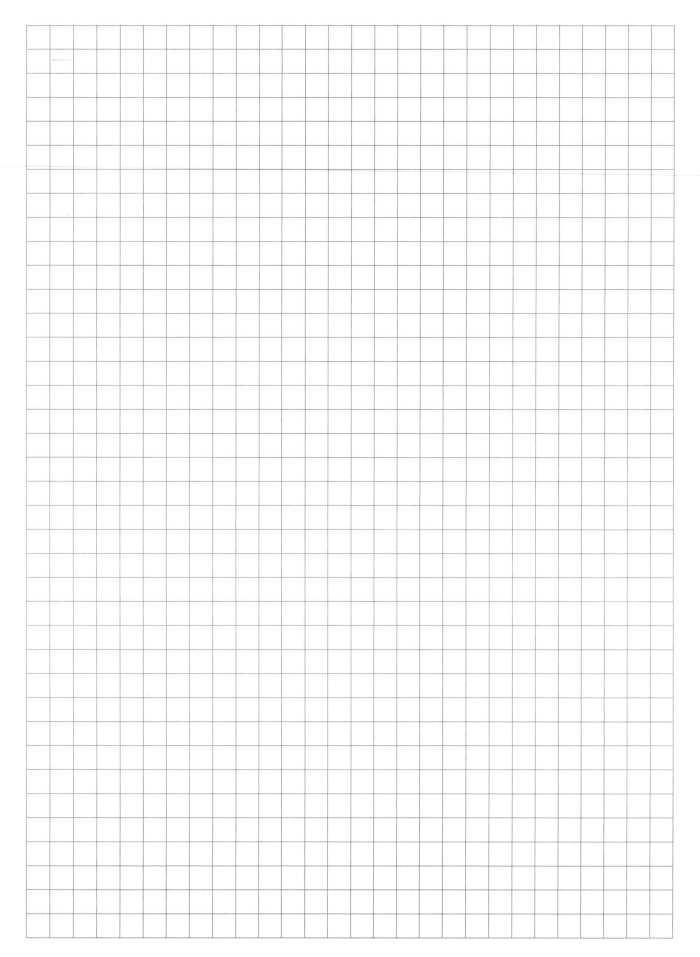

92

94

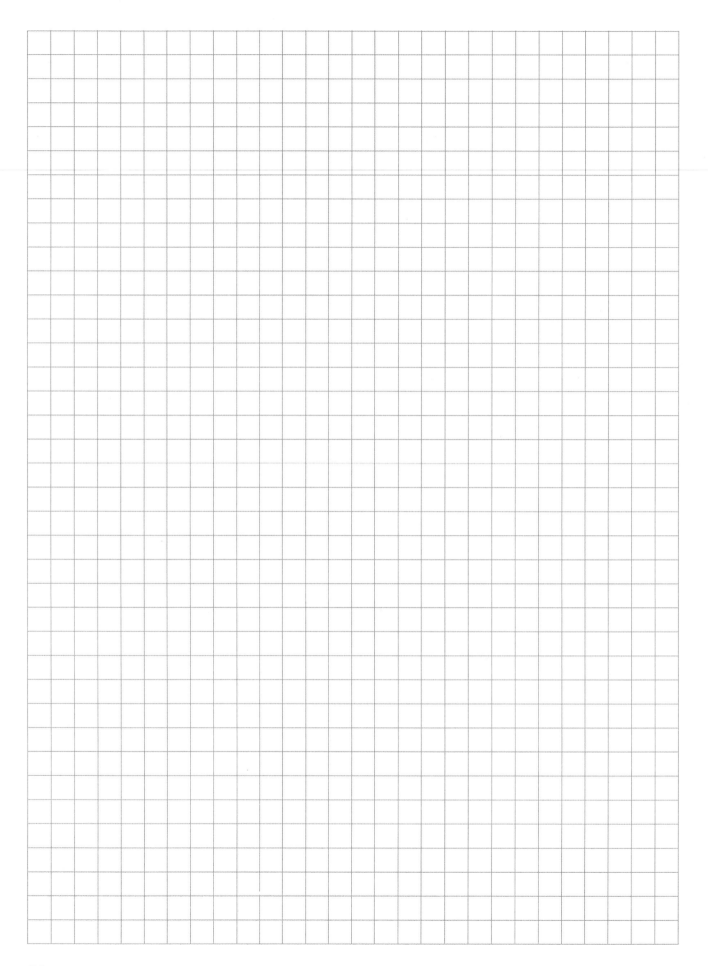

96

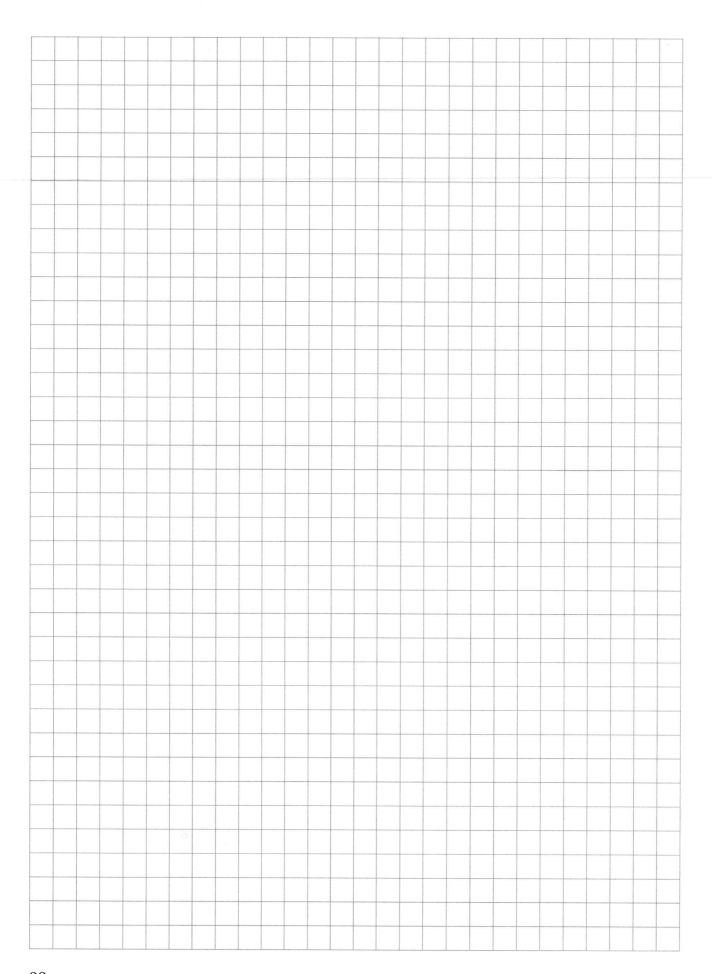

98

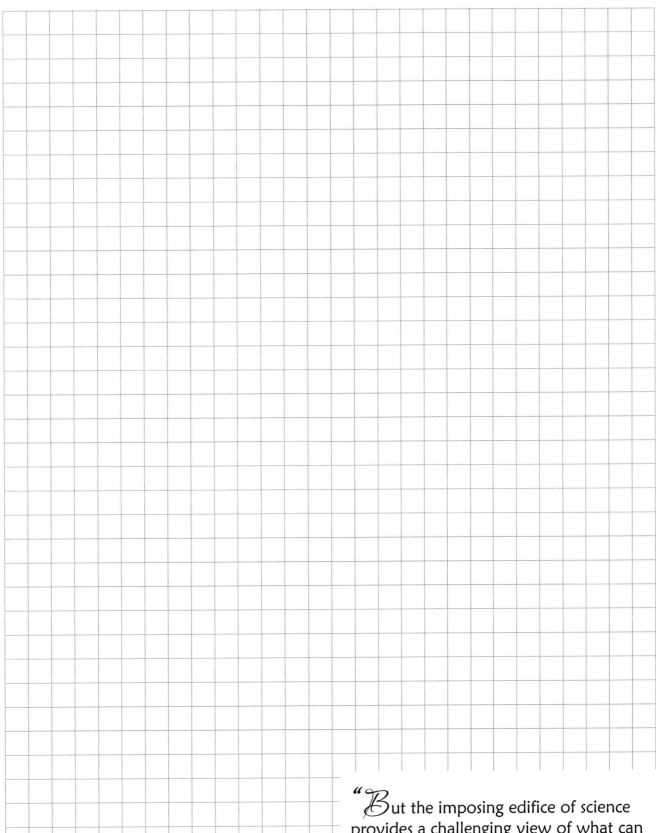

"But the imposing edifice of science provides a challenging view of what can be achieved by the accumulation of many small efforts in a steady objective and dedicated search for truth."

—Charles H. Townes

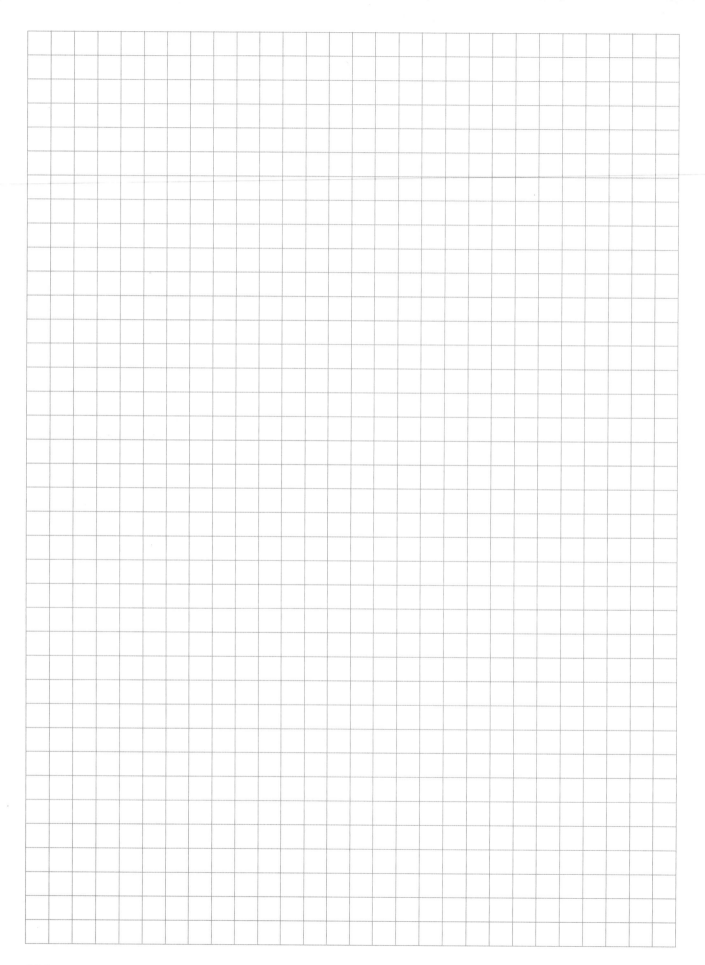

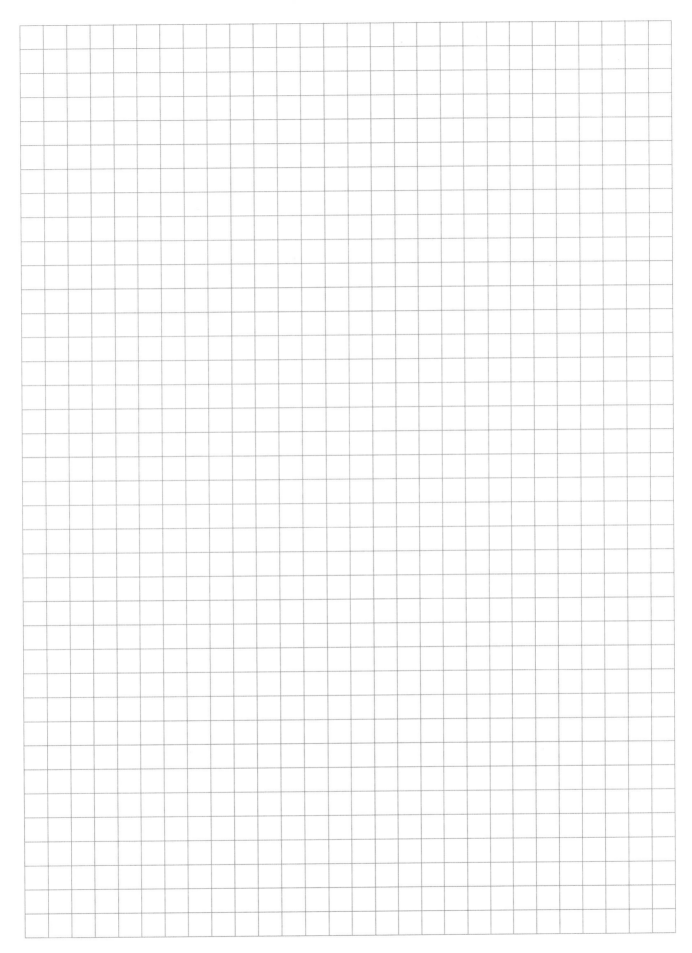

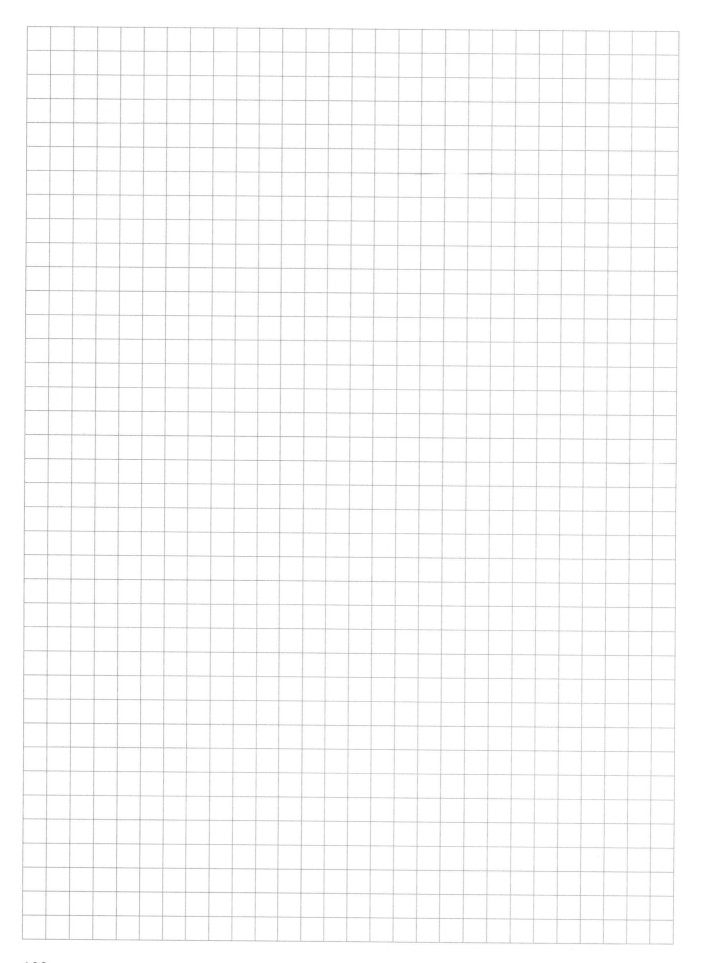

102

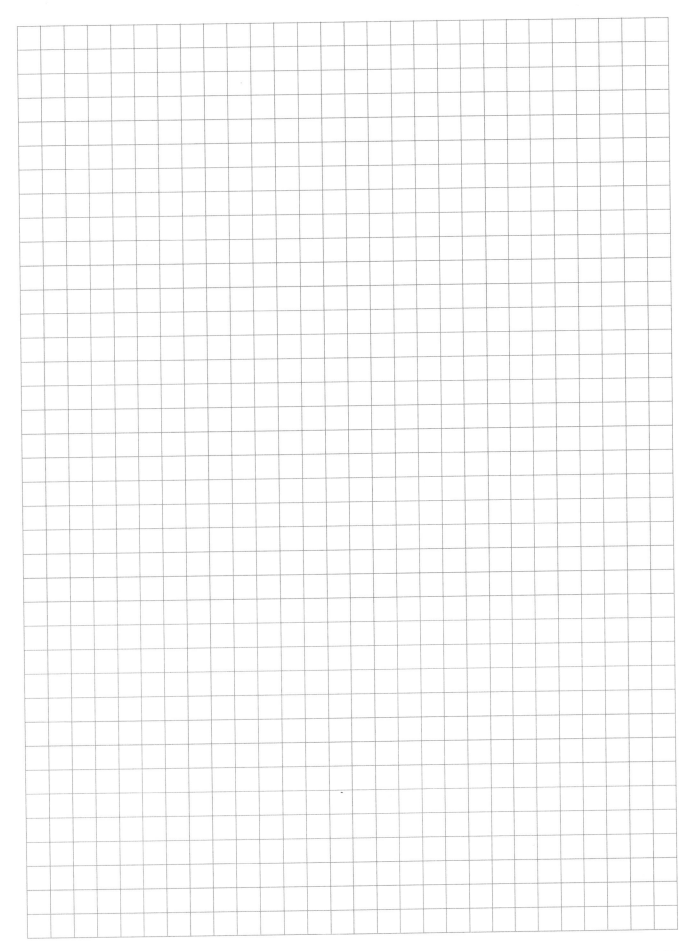

104

Reference Section

"*I am among those who think that science has great beauty.
A scientist in his laboratory is not only a technician: he is also a child
placed before natural phenomena which impress him like a fairy tale.*"
—Marie Curie

Scientific Method

Observe:

Observe or read about a phenomenon.

Hypothesize:

Wonder about your observations, and invent a hypothesis—a "guess"—that could explain the phenomenon or set of facts that you have observed.

Predict:

Use the logical consequences of your hypothesis to predict observations of new phenomena or results of new measurements.

Verify:

Perform experiments to test these predictions, to find just which prediction occurred.

Evaluate:

Search for other possible explanations of the result until you can show, with confidence, that your guess was indeed the explanation.

Publish:

Tell others of your results. Other scientists can then review your reasoning and see if they can also repeat the result. This is known as peer review.

Outline for a Lab Report

Your lab report should be divided into these sections. Be sure to include titles for each section.

I. Introduction
    - Explain the purpose of the experiment.
    - Discuss the existing knowledge on the issue (usually what you have learned in class). This gives the readers the background information they need to be able to understand your experiment and defines the presuppositions you may have.

II. Hypothesis
    - State your expected outcomes as an "if...then" statement.

III. Materials
    - List all the materials you used in the experiment including the amount of each item used.

IV. Procedure
    - Explain your process in enough detail for the reader to replicate your experiment exactly.

V. Results
    - Use charts, graphs, or other visuals to represent your results.
    - Describe any inconsistent variables that may have skewed the results.
    - Include everything you wrote down during the experiment (math, records, notes, etc.).

VI. Analysis/Discussion
    - Present your findings, noting anything that surprised you.
    - If the results are unexpected, speculate as to why.
    - Discuss anything that may have gone wrong.

VII. Conclusion
    - Affirm or deny your original hypothesis.
    - Defend this statement using the results of the experiment.

Common Ways to Present Information in Lab Reports

- **Chart/Table**: Organizes and displays data in rows and columns.
- **Diagram/Model**: Demonstrates how something works using a drawing or an example.
- **Line Graph**: Uses a line to relate two sets of data to show trends and make predictions.
- **Bar Graph**: Uses bars to show relationships between groups.
- **Pie Graph**: Divides a circle into wedges to show how parts relate to the whole.
- **Flow Chart**: Uses a series of boxes to show a sequence of events or ideas.

Parts of the Microscope

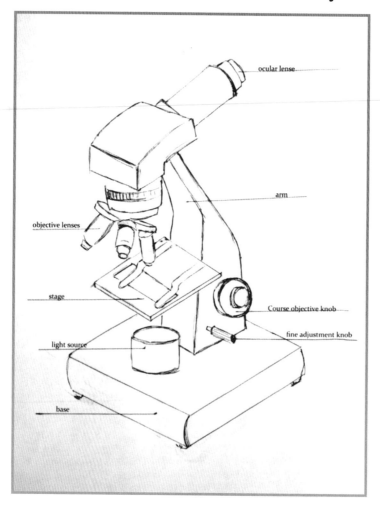

Eyepiece: Contains the ocular lens, which magnifies the image 10x.

Arm: Used to carry the microscope.

Course Adjustment Knob: The first step in focusing the specimen; moves the stage up and down for general focusing.

Fine Adjustment Knob: The second step in focusing; sharpens the focus on the specimen for greater detail.

Diaphragm: Immediately under the stage; regulates the amount of light entering through the stage and illuminating the specimen.

Base: Contains the electronics; when carrying the microscope, hold one hand below the base and the other on the arm.

Light Source: Shines light through the diaphragm, specimen, and up to the lenses.

Stage: Supports the slide; contains an opening for light to pass through directly below the specimen.

Objective Lenses: Magnifies 10x, 40x, and 100x (remember that the ocular piece adds an additional 10x magnification).

Chemistry

Selected Solubility Products and Formation Constants at 25°C

Solubility Rules: General

1. All sodium, potassium, and ammonium salts are soluble.
2. All nitrates, acetates, and perchlorates are soluble.
3. All silver, lead, and mercury(I) salts are insoluble.
4. All chlorides, bromides, and iodides are soluble.
5. All carbonates, sulfides, oxides, and hydroxides are insoluble.
6. All sulfates are soluble, except strontium sulfate and barium sulfate.

Solubility Rules: Specific

1. Salts containing Group I elements are soluble (Li^+, Na^+, K^+, Cs^+, Rb^+). Exceptions to this rule are rare. Salts containing the ammonium ion (NH_4^+) are also soluble.
2. Salts containing the nitrate ion (NO_3^-) are generally soluble.
3. Salts containing Cl^-, Br^-, I^- are generally soluble. Important exceptions to this rule are halide salts of Ag^+, Pb_2^+, and $(Hg_2)^{2+}$. Thus, $AgCl$, $PbBr_2$, and Hg_2Cl_2 are all insoluble.
4. Most silver salts are insoluble. $AgNO_3$ and $Ag(C_2H_3O_2)$ are common soluble salts of silver; virtually anything else is insoluble.
5. Most sulfate salts are soluble. Important exceptions to this rule include $BaSO_4$, $PbSO_4$, Ag_2SO_4, and $SrSO_4$.
6. Most hydroxide salts are only slightly soluble. Hydroxide salts of Group I elements are soluble. Hydroxide salts of Group II elements (Ca, Sr, and Ba) are slightly soluble. Hydroxide salts of transition metals and Al_3^+ are insoluble. Thus, $Fe(OH)_3$, $Al(OH)_3$, $Co(OH)_2$ are not soluble.
7. Most sulfides of transition metals are highly insoluble. Thus, CdS, FeS, ZnS, Ag_2S are all insoluble. Arsenic, antimony, bismuth, and lead sulfides are also insoluble.
8. Carbonates are frequently insoluble. Group II carbonates (Ca, Sr, and Ba) are insoluble. Some other insoluble carbonates include $FeCO_3$ and $PbCO_3$.
9. Chromates are frequently insoluble. Examples: $PbCrO_4$, $BaCrO_4$.
10. Phosphates are frequently insoluble. Examples: $Ca_3(PO_4)_2$, Ag_3PO_4.
11. Fluorides are frequently insoluble. Examples: BaF_2, MgF_2, PbF_2.

PERIODIC TABLE OF THE ELEMENTS

Scientific Method

WONDER

"I wonder what would happen if..."

Ask a question about an observation: how, what, when, who, which, why, or where?

ASK A QUESTION

DO BACKGROUND RESEARCH

RESEARCH

Ask lots of questions and research the topic to find out what is already known. Use the common topics of definition and authority.

CONSTRUCT A HYPOTHESIS

HYPOTHESIZE

A hypothesis is an educated guess about how things work: "If _____ [I do this] _____ , then _____ [this] _____ will happen."

TROUBLE-SHOOT PROCEDURE

TEST WITH AN EXPERIMENT

EXPERIMENT

Conduct a fair test by making sure that you change only one factor at a time while keeping all other conditions the same.

PROCEDURE WORKING?

YES

NO

ANALYZE

Analyze the data to conclude whether the hypothesis was valid. What did you learn from the experiment? What is the next step you can take with what you have learned?

ANALYZE AND DRAW CONCLUSIONS

RESULTS SUPPORT, DO NOT SUPPORT, OR PARTIALLY SUPPORT HYPOTHESIS

COMMUNICATE RESULTS

SHARE RESULTS

Document the process in scientific papers and on display boards. Scientists must be prepared to describe and discuss their findings regardless of whether their hypothesis was proven to be true.

Classical Conversations®

The Classification of Living Things

KINGDOM	Phylum	Class	Characteristics and Examples
MONERA			**Prokaryotic, one-celled, bacteria microorganisms**
	Gracilicutes		**Gram-negative (red)**
		Scotobacteria	Non-photosynthetic
		Anoxyphotobacteria	Non-oxygen-producing photosynthetic
		Oxyphotobacteria	Oxygen-producing photosynthetic
	Firmicutes		**Gram-positive (blue)**
		Firmibacteria	Bacilli or cocci (*Streptococcus*)
		Thallobacteria	Any other shape (dental and gum disease)
	Tenericutes		**Lack a cell wall**
		Mollicutes	Pneumonia causing
	Mendosicutes		**Exotic cell walls**
		Archaebacteria	Live in uninhabitable places, cyanobacteria botulism

KINGDOM	Subkingdom	Phylum	Characteristics and Examples
PROTISTA			**Eukaryotic cells, mostly one-celled microorganisms**
	Protozoa		**One-cell, mostly heterotrophic, locomotion**
		Sarcodina	Locomotion: pseudopods, ex: amoeba
		Mastigophora	Locomotion: flagellum, ex: *Euglena, Volvox*
		Sporozoa	Locomotion: none, spore producing, ex: plasmodium
		Ciliophora	Locomotion: cilia, ex: paramecium, stentor
	Algae		**Form colonies, float, mostly autotrophic**
		Chlorophyta	Fresh water, ex: desmid, *Spirogyra* (green algae)
		Chrysophyta	Marine and fresh water, ex: diatoms
		Pyrrophyta	Marine water, ex: dinoflagellate (red tide)
		Phaeophyta	Marine water, multiple cells, ex: kelp (brown algae)
		Rhodophyta	Marine water, multiple cells, ex: coral weed (red algae)

KINGDOM	Phylum	Characteristics and Examples
FUNGI		**Eukaryotic cells, multicellular, decomposers**
	Basidiomycota	"Club fungi," ex: mushrooms, puffballs, shelf fungi
	Ascomycota	"Sac fungi," ex: morel, yeast
	Zygomycota	Spores have hard covering called zygospores, ex: bread mold, fruit mold
	Chytridiomycota	Single cells, spores have flagella, ex: potato wart
	Deuteromycota	"Imperfect fungi," ex: penicillin, cheese molds
	Myxomycota	Ex: slime mold

KINGDOM	Phylum	Division	Characteristics and Examples
PLANTAE			**Eukaryotic cells, many-celled organisms, autotrophs, cell walls**
	Anthophyta		**Angiosperms (flowering and seeds), stems, leaves, roots**
		Monocotyledonae	Monocot seed, ex: lilies, rice, corn, grasses, bamboo
		Dicotyledonae	Dicot seed, largest angiosperm class, ex: magnolia, hickory, oak
	Coniferophyta		**Gymnosperms (naked seed) evergreen, cones, needles, scales**
		Pinopsida	Ex: pines, cedars, hemlocks, cypress, juniper, larches, firs, and yews
	Lycopodiophyta		**Spores, green, branched stems, small scale-like leaves, and rhizomes**
		Lycopodiopsida	Ex: club mosses, quillworts, and spike mosses
	Pteridophyta		**Spores, vascular, leaves, roots, and stems; do not have flowers**
		Filicopsida	Ex: fiddlehead and all ferns
	Bryophyta		**Spores, non-vascular, true mosses (95% of all mosses)**
		Bryopsida	Ex: peat, sphagnum, copper mosses

KINGDOM	Phylum	Class	Characteristics and Examples
ANIMALIA			**Eukaryotic cells, many-celled organisms, heterotrophs**
	Chordata		**Animals with a spinal cord**
		Mammalia	Warm blooded, hair, live young, mammary glands, ex: man, horse
		Aves	Warm blooded, feathers, eggs, wings, hollow bones, ex: eagle, duck
		Agnatha	Cold-blooded, fish without jaws, ex: lamprey
		Chondrichthyes	Cold blooded; cartilage fish, ex: shark
		Osteichthyes	Cold blooded, bony fish, ex: trout
		Reptilia	Cold blooded, dry scales, leathery eggs, lungs, ex: turtle, crocodile
		Amphibia	Cold blooded, moist thin skin, eggs, gills and lungs, ex: frog, newt
	Arthropoda		**Largest phylum, exoskeleton, jointed limbs**
		Arachnida	Ex: spiders
		Crustacea	Ex: crabs
		Diplopoda	Ex: millipedes
		Chilopoda	Ex: centipedes
	Echinodermata		**Sea stars and relatives**
		Asteroidea	Ex: sea star or starfish
		Ophiuroidea	Ex: brittle stars, basket stars, serpent stars
		Echinoidea	Ex: sea urchins, heart urchins, sand dollars
		Holothuroidea	Ex: holothurians or sea cucumbers
		Crinoidea	Ex: feather stars, sea lilies
	Mollusca		**Mollusks: soft bodies, hard shell**
		Gastropoda	Ex: snails, slugs, conchs, whelks, cowries
		Bivalvia	Ex: clams, mussels, oysters, cockles, scallops
		Polyplacophora	Ex: chitons
		Scaphopoda	Ex: tusk shells
		Cephalopoda	Ex: octopus, squid, cuttlefish, nautilus
	Annelida		**Segmented worms**
		Oligochaeta	Ex: freshwater worms, earthworms
		Polychaeta	Ex: marine worms
		Hirudinea	Ex: leeches
	Nematoda		**Roundworms: often microscopic, bilateral symmetry**
		Adenophorea	Ex: *Trichinella*
		Secernentea	Ex: *Ascaris*
	Platyhelminthes		**Flatworms: bilateral symmetry**
		Trematoda	Ex: fluke
		Turbellaria	Ex: *Planaria*
		Cestoda	Ex: tapeworms
		Monogenea	Ex: flukes that have one host for a lifetime
	Porifera		**Sponges: immobile, asymetrical, feed by filtering water**
		Demospongia	Ex: spongin - most diverse, 90% of sponges
		Calcarea	Ex: calcium
		Hexactinellida	Spicules of silica ex: glass sponges
	Cnidaria		**Stinging - cell animals, radial symmetry**
		Anthozoa	Ex: corals, sea anemones
		Scyphozoa	Ex: swimming jellyfish
		Staurozoa	Ex: stalked jellyfish
		Cubozoa	Ex: box jellyfish
		Hydrozoa	Ex: siphonophores and hydroids

Timeline
of Famous Scientists

400 BC

HIPPOCRATES began the study of medicine.

287 BC

ARCHIMEDES invented the sciences of mechanics and hydrostatics and discovered the laws of levers and pulleys.

100 BC

HIPPARCHUS cataloged the stars.

1 AD

PTOLEMY formed the geocentric theory of the solar system.

LEONARDO DA VINCI developed scientific study through observation (similar to Scientific Method).

COPERNICUS formed the heliocentric view of the solar system.

1400

"Finally I give this advice to any people who might be completely unfamiliar with mathematical questions... If such readers were to be terrified by the difficulty of the geometrical argument, they might deprive themselves of the most joyful fruit of contemplating the harmonies. Now, let us go to work with God." —Kepler

1500

GALILEO GALILEI invented the telescope.

JOHANNES KEPLER formulated three laws of planetary motion.

1600

BLAISE PASCAL invented the calculator.

"Let man then contemplate the whole of nature in her full and grand majesty, and turn his vision from the low objects which surround him. Let him gaze on that brilliant light, set like an eternal lamp to illumine the universe... In short, it is the greatest sensible mark of the almighty power of God, that imagination loses itself in that thought." —Pascal

ROBERT BOYLE launched the field of chemistry.

ISAAC NEWTON developed laws of universal gravitation and motion.

1700

"[When] I study the book of nature, I find myself oftentimes reduced to exclaim with the Psalmist, How manifold are Thy works, O Lord! in wisdom hast Thou made them all!" —Boyle

DANIEL BERNOULLI developed principles of the flow of fluids.

"Whence is it that Nature doth nothing in vain; and whence arises all that Order and Beauty which we see in the World? ... does it not appear from Phænomena that there is a Being incorporeal, living, intelligent, omnipresent..." —Newton

CARL LINNAEUS developed the precursor to the modern classification system of living things.

"...[man] is noble in his nature, in as much as, by the powers of his mind, he is able to reason justly upon whatever discovers itself to his senses; and to look, with reverence and wonder, upon the works of Him who created all things....It is therefore the business of a thinking being, to look forward to the purposes of all things; and to remember that the end of creation is, that God may be glorified in all his works."

1800

MICHAEL FARADAY invented the electric motor.

"Shall we educate ourselves in what is known, and then casting away all we have acquired, turn to our ignorance for aid to guide us among the unknown? If so, instruct a man to write, but employ one who is unacquainted with letters to read that which is written; the end will be just as unsatisfactory, though not so injurious; for the book of nature, which we have to read, is written by the finger of God." ("On Mental Education" lecture, 1854)

GREGOR MENDEL formalized the study of genetics.

"So natural and supernatural must unite to the realization of the Holiness to the people. Man must contribute his minimum work of toil, and God gives the growth. Truly, the seed, the talent, the grace of God is there, and man has simply to work, take the seeds to bring them to the bankers. So that we 'may have life, and abundantly.'" ("Sermon on Easter," c. 1867)

... I have the capacity of being more wicked than any example that man could set me, and ... if I escape, it is only by God's grace helping me to get rid of myself, partially in science, more completely in society, —but not perfectly except by committing myself to God." (Letter to Rev. C. B. Tayler)

JAMES CLERK MAXWELL discovered the relationship between electricity and magnetism.

THOMAS EDISON invented the light bulb, phonograph, and motion picture camera.

JOSEPH LISTER developed antiseptic surgery.

MAX PLANCK originated quantum theory.

"Both religion and natural science require a belief in God for their activities, to the former He is the starting point, to the latter the goal of every thought process. To the former He is the foundation, to the latter, the crown of the edifice of every generalized world view." ("Religion and Natural Science," lecture 1937)

ALBERT EINSTEIN developed the theory of relativity.

1920

NEILS BOHR developed the Bohr model of the atom.

EDWIN HUBBLE discovered galaxies beyond the Milky Way.

1950

JAMES WATSON & FRANCIS CRICK published the structure of the DNA molecule.

1900

SAMUEL MORSE invented the telegraph and Morse code.

CHARLES DARWIN popularized the theory of natural selection and the theory of evolution.

CHARLES BABBAGE invented a calculating engine and became the father of computing.

"Almost all thinking men who have studied the laws which govern the animate and the inanimate world around us, agree that the belief in the existence of one Supreme Creator, possessed of infinite wisdom and power, is open to far less difficulties than the supposition of the absence of any cause, or the existence of a plurality of causes." (Passages from the Life of a Philosopher)

LOUIS PASTEUR confirmed the germ theory of disease and developed pasteurization.

"Absolute faith in God and in Eternity, and a conviction that the power for good given to us in this world will be continued beyond it, were feelings which pervaded his whole life" (Vallery-Radot [son-in-law], The Life of Pasteur, 1911).

LORD KELVIN developed an absolute temperature scale and formulated the second law of thermodynamics.

"But overpoweringly strong proofs of intelligent and benevolent design lie all round us, and if ever perplexities, whether metaphysical or scientific, turn us away from them for a time, they come back upon us with irresistible force, showing to us through nature the influence of a free will, and teaching us that all living beings depend on one ever-acting Creator and Ruler." (Presidential Address to the British Association for the Advancement of Science, 1871)

MARIE CURIE discovered polonium and radium.

GEORGE WASHINGTON CARVER used chemistry to improve agricultural production in the American South.

"How I thank God every day that I can walk and talk with Him. Just last week I was reminded of His omnipotence, majesty and power through a little specimen of mineral sent me for analysis from Bakersfield, California. I have dissolved it, purified it, made conditions favorable for the formation of crystals, when lo before my very eyes, a beautiful bunch of sea green crystals have formed and alongside of them a bunch of snow white ones. Marvel of marvels, how I wish I had you in God's little workshop for a while, how your soul would be thrilled and lifted up." (1927 letter to YMCA official Jack Boyd)

Measurements

Quantity	Base Unit - Metric	Base Unit - English
Mass	gram (g)	slug (sl)
Distance	meter (m)	foot (ft)
Volume	liter (L)	gallon (g)
Time	seconds (s)	second (s)
Temperature	Celsius (C)	Fahrenheit (F)
Electric current	ampere (A)	ampere (A)
Amt. of matter	mole (mol)	mole (mol)

Prefix (Symbol)	Numerical Meaning
tera- (T)	10^{12} = 1,000,000,000,000
giga- (G)	10^{9} = 1,000,000,000
mega- (M)	10^{6} = 1,000,000
kilo (k)	10^{3} = 1,000
hecto- (h)	10^{2} = 100
deca- (da)	10 = 10
deci- (d)	10^{-1} = 0.1
centi- (c)	10^{-2} = 0.01
milli- (m)	10^{-3} = 0.001
micro- (μ)	10^{-6} = 0.000001
nano- (n)	10^{-9} = 0.000000001

Converting Between U.S. Customary Units

Length:
1 inch = 1/12 foot
1 foot = 12 inches or 1/3 yard
1 yard = 36 inches or 3 feet
1 mile = 5,280 feet or 1,760 yards

Liquid Volume or Capacity:
1 fluid ounce = 1/16 pint
1 pint = 16 fluid ounces or 2 cups
1 quart = 2 pints or 1/4 gallon
1 gallon = 4 quarts or 8 pints or 16 cups

Weight:
1 pound = 16 ounces
1 ton = 2,000 pounds

Converting from U.S. Customary Units to Metric Units
1 inch = 2.54 centimeters
1 inch = 25.4 millimeters
1 foot = 30.48 centimeters
1 yard = 0.91 meters
1 mile = 1.61 kilometers
1 fluid ounce = 29.57 milliliters
1 pint = 0.47 liters
1 quart = 0.95 liters
1 gallon = 3.79 liters
1 ounce = 28.35 grams
1 pound = 0.45 kilograms

Converting Between Celsius and Fahrenheit		
$F = (C \times 9/5) + 32$		
$C = (F - 32) \times 5/9$		
Boiling Point of Water	212°F	100°C
Normal Body Temperature	98.6°F	37°C
Freezing Point of Water	32°F	0°C

Parts per million (ppm): The number of molecules (or atoms) of a substance in a mixture for every 1 million molecules (or atoms) in that mixture (1% = 10,000 ppm)

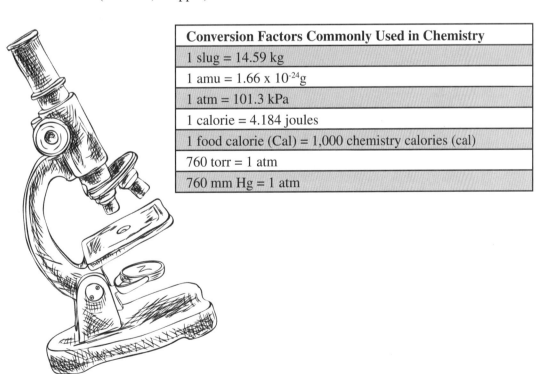

Conversion Factors Commonly Used in Chemistry
1 slug = 14.59 kg
1 amu = 1.66×10^{-24}g
1 atm = 101.3 kPa
1 calorie = 4.184 joules
1 food calorie (Cal) = 1,000 chemistry calories (cal)
760 torr = 1 atm
760 mm Hg = 1 atm

Common Equations

Density = mass/volume $d = \frac{m}{v}$

Force = (mass) x (acceleration)   F = (m)(a)

Speed = $\frac{distance\ traveled}{time\ traveled}$

Acceleration = $\frac{final\ velocity - initial\ velocity}{time}$

Temperature change = (mass)(specific heat)(change in temperature)   q=(m)(c) T

Frequency of a light wave = $\frac{speed\ of\ light}{wavelength}$

Energy of a light wave = $\frac{Planck's\ Constant}{frequency}$

Planck's Constant (h) = 6.63×10^{-34} joule-seconds

Plain and Solid Geometry

Circle
Area = πr^2
Circumference = $2\pi r$

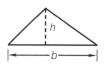

Triangle
Area = $\frac{1}{2} bh$

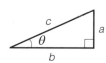

Right Triangle
Pythagorean theorem: $c^2 = a^2 + b^2$

$sin\ \theta = \frac{opposite}{hypotenuse} = \frac{a}{c}$

$cos\ \theta = \frac{adjacent}{hypotenuse} = \frac{b}{c}$

$tan\ \theta = \frac{sin\ \theta}{cos\ \theta} = \frac{opposite}{adjacent} = \frac{a}{b}$

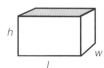

Right Rectangular Solid
Volume = $l \times w \times h$

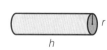

Right Circular Cylinder
Volume = $\pi r^2 h$

Sphere
Volume = $\frac{4}{3} \pi r^3$
Surface Area = $4\pi r^2$

Quadratic Equation

$$ax^2 + bx + c = 0 \qquad x = \frac{-b \pm \sqrt{b^2 - 4ac}}{2a} \qquad (a = 0)$$

Moments of Inertia

Cylindrical Hoop $I = mr^2$	**Solid Cylinder or Disk** $I = \frac{1}{2} mr^2$	**Solid Sphere** $I = \frac{2}{5} mr^2$
Hollow Sphere $I = \frac{2}{3} mr^2$	**Long, Thin Rod About Center** $I = \frac{1}{12} mL^2$	**Long, Thin Rod About End** $I = \frac{1}{3} mL^2$

Physics

Density = Mass / Volume

Pressure = Force / Area

Work = Force × Displacement

Speed = Distance / Time

Average speed = Total Distance / Total Time

Velocity = Displacement / Time

Acceleration = Velocity / Time

Force = Mass × Acceleration

Power = Work / Time

Centripetal Force = Mass × Velocity^2 / Radius

Centripetal Acceleration = Velocity^2 / Radius

Human Organ Systems

System	Structures	Functions
1. Skeletal System	• bones • cartilage • ligaments • tendons	1. Supports the body and protects internal organs 2. Blood cell formation
2. Muscular System	• skeletal muscle • smooth muscle • cardiac muscle	1. Works with skeletal system to produce movement 2. Helps circulate blood and food
3. Circulatory System	• heart • blood vessels • blood	1. Transports oxygen, carbon dioxide, food, and hormones 2. Maintains constant temperatures 3. Fights disease
4. Digestive System	• mouth • esophagus • stomach • small intestine • large intestine • rectum • pancreas • liver	1. Converts food into smaller molecules 2. Eliminates waste
5. Excretory System	• skin • lungs • kidneys • ureters • urinary bladder • urethra	1. Eliminates waste products from the body
6. Respiratory System	• nose • pharynx • larynx • trachea • bronchi • bronchioles • lungs	1. Exchange of oxygen gas and the removal of carbon dioxide
7. Nervous System	• brain • spinal cord • peripheral nerves	1. Receives and transmits information in the body

CLASSICAL CONVERSATIONS.COM

System	Structures	Functions
8. Endocrine System	• hypothalamus • pituitary • thyroid • parathyroids • adrenals • pancreas • ovaries • testes	1. Regulates overall metabolism, growth, reproduction, and maintenance of homeostasis
9. Reproductive System	• testes • epididymis • vas deferens • penis • ovaries • fallopian tubes • uterus • vagina	1. Produces reproductive cells 2. Development of the baby
10. Lymphatic/Immune System	• white blood cells • thymus • spleen • lymph nodes • lymph vessels	1. Helps protect the body from disease 2. Collects and returns fluid to the circulatory system
11. Integumentary System	• skin • hair • nails • sweat glands • oil glands	1. Protects against infection, UV radiation, and injuries 2. Helps regulate body temperature

Classical Acts and Facts® Science Cards put science right at your fingertips!

Marie Curie once said, "I am among those who think that science has great beauty." With the beautiful images and artwork that accompany each clearly-explained science fact, you will come to believe in the beauty of science, too. The high-quality, laminated 5″ x 8″ cards are arranged into four sets according to major science categories. This exclusive series of science cards is designed to serve you throughout your educational journey, with a comprehensive set of grammar pegs that relate to Foundations memory work all the way through the Challenge program's sciences. Over 120 unique acts and facts in all!

Biology and Geology
Ecology, Astronomy, Physics
Anatomy, Chemistry, Origins
Famous Scientists and More

Nature Sketch Journal

Inspire your students to develop their observation skills through drawing and recording research in this beautiful sketch journal developed by Classical Conversations parents. This book alternates pages for drawing and pages for writing and sprinkles inspirational quotes throughout for students to discover as they fill in the pages. A reference section in the back includes interesting science facts, a classification of living things chart, a timeline of famous scientists, the periodic table, and measurements and conversions tables. A ruler on the back cover makes it easy to make measurements in the classroom or on a nature hike.

Discovering Atomos

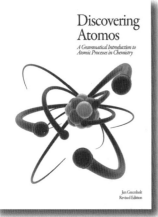

This booklet is designed to give 6th–8th graders a basic introduction to chemical processes at the atomic level. The text is divided into six simple lessons that introduce various topics and then provide examples and exercises to support the new concepts. *Discovering Atomos* is a consumable workbook. Allowing students to write directly in the manual makes it easy for them to look back at helpful charts and figures, compare their work to the examples given, and reference the information that they have learned. The lessons in the text are meant to create a foundation for further studies in chemistry. If students can memorize the periodic table and learn how to combine elements now, higher level chemistry will be less intimidating right from the start. If students can learn that understanding chemistry is feasible and even fun, their success in the subject will already be well on its way. *Atomos* is a Greek word from which the word "atom" is derived. *Atomos* means "indivisible" (*a* means "not"; *temnein* means "to cut").